Forewords

"In the beginning, God created the heavens and the earth."
(World Messianic Bible, Genesis)

In the Galaxy, billions of stars are twinkling and assemble into the Milky Way at the night sky. Constellations tell us an ancient story in the Greek Mythology. A majority of stars that we see in the sky is born, grown up, and gone in the Galaxy. The sun is one of the stars. Stars increase their brightness after the birth, and bloom a big flower of fireworks in space. Some stars foster living things on their planets. Thereafter, the stars are getting old and finish their lives.

We could feel various charms of the beauties of nature upon thinking about the pretty and transitory spiral flower of the Galaxy. We are born and grown in the prime together with other living things on the earth, which is one of planets of the solar system in the Galaxy. All the nature and human societies in front of us are also created, grown up, and bloom as a part of the Galaxy. When we see the beauty of them, we could not prevent feeling awe to the mighty power of God for creation of the heavens and the earth. We wonder how such an amazing world has been created in order.

Astronomy clearly tells us without doubt that the Sun

was born as a member of the Galaxy. However, we still have lots of questions on the Galaxy. When, where, and how was the Sun created in the Galaxy? And how was the mother Galaxy was born in the vast and boundless whole world or the cosmos? Astronomers suggest that the Sun has its life span. Then, does the Galaxy also have its life span? Furthermore, do other many galaxies in the cosmos also have their life spans and destinations as of the Galaxy? Finally, does the whole cosmos have its birth and the end?

Since scientists still argue on the mechanism of the birth and the end of the Sun, a life story of the Galaxy should be quite uncertain. However, we have many astronomic evidences unique to the Galaxy but not to the solar system, which may give us a cue of a life story of the Galaxy. It is my great pleasure to establish some scientific evaluation of the valuable human life from the point standing on the astronomy of the life of the Galaxy, which should be concluded from rigid scientific data observed by astronomers. That is the aim of this book.

At home in Kasukabe
Jun 29th, 2015
Hiroyuki Aizawa

Contents

Chapter One
Species of the Galaxy

A lot of stars twinkle brilliantly in the night sky on a fine day.   Stars do not distribute evenly in the whole sky but gather in some regions and sparse in others. Especially, it appears that a big star river flows at the night, which we call the Milky Way because it appears like milk running from the breast.   In Japanese, we call it "the River in Heaven".   A famous Japanese poet, Basho, composed as follows:

"The Galaxy, stretching over the Sado isle in the turbulent sea."

In an ancient china myth, a cowboy named Altair could enjoy a secret meeting with a weaving lady named Vega on July 7th only once a year when a bridge across the Galaxy was opened.   The great work of art created by stars in the Galaxy captivates all the people in the western and eastern world, who called it figuratively milk and river in the heaven, respectively.

A high-resolution telescope clearly shows many strange luminous bodies with a spiral shape among the stars in the night sky.   The whirlpool is an aggregation of stars called a galaxy.   Our solar system belongs to one of galaxies, the Galaxy, which is also called the Milky Way from its shape when observed from inside.

How many types of galaxy do we have in the cosmos? We see a large number of galaxies in the cosmos, and there are no same galaxies among them. Each galaxy presents its unique shape. The size is different from one to another. The apparent brightness is quite different one by one. The number of stars is also different among galaxies, while we could count it in not all the galaxies. In early twentieth century, an American astronomer Dr. Hubble classified galaxies into three types based on the shape and photonic data obtained by the telescope with the highest resolution at that period. Based on the classification, Dr. Hubble established an evolutional sequence of galaxy. According to his theory, galaxies are largely classified into two types, that is, an elliptic galaxy and a spiral galaxy. The latter is further classified into a normal or a barred spiral. Galaxies belonging to one of the three types are orderly aligned in a sequence within each type according to their axes ratio or flattening. We also find some irregular galaxies, which do not show any obvious symmetric structures and consequently do not belong to any one of the three types of Hubble sequence.

Let us learn about each type of galaxy in the Hubble sequence, especially on their structure, motion, distribution of elements, and optical nature, in detail from now on.

Chapter Two
Structure of the Galaxy

An elliptic galaxy is a luminance body with a circular or elliptic shape without individual stars. The brightness of the galaxy positively correlates to the distance from the center of the body. A statistical analysis of a huge number of elliptic galaxies clearly revealed that the three dimensional shape of an elliptic galaxy is sphere or a convex lens with the axial ratio distributing from one to a quarter. There are no elliptic galaxies with an axial ratio less than a quarter. Since all the galaxies with an axial ratio of less than a quarter belong to the spiral galaxy, it is proposed that an elliptic galaxy is an immature type of the spiral galaxy. In other words, a galaxy may be born as a spherical luminous body, and the body grows and extends gradually within a plane to form a convex lens shape, and further extends spiral arms from the edge of the lens to become a spiral galaxy during its maturation. Because we could not detect any sign of the existence of stars within an elliptic galaxy even by a modern telescope with the highest resolution, it is possible that an elliptic galaxy is at a pregnant stage with baby stars.

A spiral galaxy is an aggregate of stars, which form several spiral arms extending from the center of the galaxy. It appears just like a hurricane in the Atlantic Ocean or a typhoon in the Pacific Ocean. Based on a structure of the central part of the spiral, spiral galaxies are classified into two types, that is, a normal spiral galaxy and a bar-shaped one. The central part of the normal spiral galaxy is a luminous body with a convex lens-like shape just like the elliptic galaxy, in which we could not detect any sign of stars. We could not detect stars in relatively short arms, either, which extend from and twine tightly around the lens body. On the other hand, we could detect stars in the extended spiral arms, which expand sparsely to each other with relatively small central lens body. Even in those spirals, we could not see stars in the central body and in arms close to the central body. In a well-expanded spiral with little lens body at the center, we could see a lot of stars all along the spiral arms, which extend well and expand sparsely like blooming in the cosmos. In a spiral galaxy, the number of arms is mostly two, and sometimes becomes three, four, and could be more.

A bar-shaped spiral galaxy is basically similar to that of a normal one in its structure, except for the bar-shaped central body rather than the lens-shaped one. Usually two large spiral arms are extended from ends of the bar, while a few additional arms could extend occasionally from the bar in the spiral. The arms in the bar-shaped spiral tend to expand more sparsely than those in the normal spiral. Just like a normal spiral, a bar-shaped spiral forms a lot of stars in the outer portion of the arms but not so much in the inner portion of the arms. And the central bar does not contain any stars in it. Taken all together, a bar-shaped spiral shows quite similar aspects in its structure to a normal one except for the bar-like structure of the central luminous body.

An elliptic galaxy appears quite similar to the central luminous body of a spiral galaxy when judged from its structure and shape. Dr. Hubble suggested the possibility that an elliptic galaxy evolves to become a spiral galaxy during its maturation. There is a significant positive correlation between the extension of spiral arms and star formation. In an immature spiral, short arms have twined tightly around the central luminous lens just like a bud before blooming. In such a bud spiral, stars are not born yet while clouds in the lens body and short arms are radiating light. On the other hand, a matured and blooming spiral galaxy has a small lens with long extending arms, in which a lot of stars are twinkling. It looks like fireworks in the night sky. Dr. Hubble proposed a model of galactic evolution just like fireworks. We will come back to verify the hypothesis on a galactic evolution in chapter six after careful inspection of scientific data of the motion, elements, and optics of galaxies.

Chapter Three
Motion of the Galaxy

In the book "The Realm of the Nebulae", Dr. Hubble wrote that the sun traveled six hundred kilometer in a second, while its direction was not rigidly determined. One observation suggested that the sun travels toward the Horned Goat while another suggested that the sun travels toward the North Star. Recent astronomic observations revealed that almost all the stars move at a constant speed of around two hundred kilometer per second on the tangent line of spiral arms in the Galaxy. In other words, the Galaxy swirls as a tornado or a hurricane in the cosmos. The speed of the wind is almost constant everywhere in the cosmic tornado except for the central body, in which the speed of rotation positively correlates to the distance from the center of the vortex. Consequently, the center of the vortex does not appear to rotate.

Although the speed and direction of stars in the Galaxy has been determined, we still do not know the force applying to stars. Thus, we do not know whether the sun is falling down into the center of the Galaxy or the sun is releasing outward from the Galaxy. Modern astronomers prefer the former possibility while Dr. Hubble suggested the latter. In order to address this question, we had better take a new strategy, that is, building up several possible models of the force that regulates the motion of the Galaxy.

Then, we can inspect each model by careful examination of the aspects speculated from the model with scientific data observed by modern astronomers.

The first model is the universal gravity model. In this model, the Galaxy incorporates all materials around it into the center by universal gravity. Some astronomers call the center "black hole". It is a good idea to remind a water flow in a sink as a two-dimensional example of this model. Water flows to a drain in a sink according to the force of gravity. Usually, we see a vortex of water around the drain. Thus, it is possible for the Galaxy to perform a vortex movement around its center when materials accumulate by universal gravity in space. Let us speculate the motion of the Galaxy based on the universal gravity model. There might be gasses and small cosmic dusts in space around the Galaxy. Those materials will start to accumulate into a center of the gravity. They will assemble slowly at first, and gradually faster and faster into the center of the spiral. During the accumulation, the density of materials increases at the center of the gravity. Accordingly, the speed of accumulation will increase gradually.

During the accumulation, behavior of the materials depends on their states, that is, gas, liquid, and solid states. Solid cosmic dusts, which distribute in vast space, will combine with each other to make a large aggregate by accidental collisions or by the universal gravity. The aggregate incorporates more cosmic dusts around and grows faster by stronger universal gravity as it becomes bigger and heavier. In this way, cosmic dusts grow to become a meteorite or a star, which is attracted to the center of the Galaxy. A free falling solid body travels much faster as it falls closer to the center of the gravity. Since the pressure does not affect the volume of a solid body so much as that of gas, all the accumulated cosmic dusts should be assembled to form a huge spherical body at a center of the Galaxy. It should be noted that the universal gravity is an attractive force between two bodies along a line passing through the two. Thus, the universal gravity could drive solid bodies in an elliptic motion, which includes a straight or circular motion as a special type. And the universal gravity could not cause a spiral motion of solid bodies in space by itself. However, frictional resistance of media against the rotating solid body decreases the radius of the ellipse until the body falls down to the center of the spiral. Consequently, all the solid body should receive some frictional resistance from the medial fluid in order to swirl around the center of mass by the universal gravity. And the accumulated huge mass will form a black hole, which absorbs everything around it.

It is also possible for the media fluid to swirl around the center of mass by the universal gravity just like a spiral flow of water around the drain in the sink. In space, thin gas might swirl around the center of mass during its condensation by the universal gravity. Thin gas increases in the temperature at the center of the spiral because this process is an adiabatic compression. The compressed gas quickly increases in pressure. The high pressure might cause repulsion of the gasses to start expansion under the law of the ideal gas. Further compression results in liquefaction and solidification of the gas. Since temperature of materials does not affect on the universal gravity, liquefied and solidified materials will be fall down into the center of the spiral by the gravitation.

The second model is the space tornado model. In this model, a cosmic wind of thin gasses blows to create space tornados as a motive force for the Galaxy to swirl. According to the ideal gas law, gasses expand and shrink when it becomes hot and cool, respectively. If the two phenomena occur here and there in space, cosmic winds should blow from the high atmospheric pressure area to the low ones. The cosmic wind may blow in a different direction and at a different speed from those of the neighboring atmospheres. Where fluid media path through each other, whirlwinds occur between them. For example, a whirlwind or a dust devil occurs when a wind blows on the earth. It is quite likely that thin gasses may swirl in space in the same way as on the earth. The source of the power to create a space tornado should be the spatial and temporal fluctuation of an internal energy and pressure of the space atmosphere. In other words, the source of the power is the breathing of the cosmos.

In the cosmos, gasses flow at a speed of the space wind, which swirls here and there streaming into a center of the spiral. Consequently, the speed of gasses at each point in the spiral should be almost equal to that of the wind and does not depend on the distance from the center except for the central region of the tornado. On the other hand, the speed of the wind decreases at the center of the spiral, and the streamed gasses start to shoot upward or downward from the center of the spiral along its rotational axis. Thus, the gasses near the central region are accelerated upward and downward. The gas is compressed as it flows into the center of the spiral, and the pressure becomes max at the center. Because of the diabetic compression of the gas at the center of the spiral, the temperature of the gas should be the highest there. In the tornado, cosmic dusts in the space wind may also stream into the spiral, and assemble into large aggregates by the high pressure of a tornado. Small and tiny aggregates of cosmic dusts may be released from the tornado on the upward or downward flow of gasses. Also, a tornado may strew large dusts, meteorites, and stars radially in a spiral plane by the centrifugal force. How is the behavior of those strewed bodies such as meteorites and stars? Let us spin a wet umbrella in a rainy day, and we will find that raindrops flow away radially on a straight line by the centrifugal force. In the same way, meteorites and stars may flow away radially on a straight line from the center of the space tornado. If gasses still keep on pouring into the tornado in a spiral manner, the radially travelling cosmic dusts, meteorites, and stars may flow away swirling on the spiral stream of the gas.

In the tornado hypothesis, the Galaxy is formed by combination of six different types of motions of materials, that is, (1) the gaseous flow swirling into the center, (2) the dust flow swirling into the center on the gaseous flow, (3) the escaping gaseous flow from the center of the spiral upwardly and downwardly along the spiral axis, (4) the cosmic dusts flow from the center of the spiral upwardly and downwardly along the spiral axis, and (5) the flow of large dusts and meteorites strewing around radially from the center within a spiral plane by the centrifugal force, and (6) the rotational movement of the radially released materials after riding on the whirlwind.

The third model is the rotating boiler model. In this model, the Galaxy is created from the high-temperature and high-pressure vapors shooting from a steam engine, which is rotating just like a "wind ball" of the ancient Alexandria. The rotating steam engine spouts out several lines of high-temperature and high-pressure vapors, which expand into the cosmic space and consequently condense into liquid droplets after radiation cooling. In this model, the spouting vapors travels away radially on a straight line. Since the spouting body is rotating, the clouds of the vapor naturally form a spiral shape without any rotation around the center.

The vapor spouts out from gas ports of the wind ball, and condenses by diabetic expansion and radiation cooling to create stars. In this hypothesis, stars will be created at outer regions of a disk in the Galaxy because the vapor is expected to condense after expansion and radiation cooling. The baby stars travel radially leaving the steam engine far behind. In the rotating boiler model, the mechanism of birth and movement of stars in the Galaxy is quite simple. The spiral of the Galaxy is just the shape and not the motion of materials. In fact, vapors are spouted out from nozzles of rotating wind ball and the vapors travel away radially to form a spiral shape. The most difficult point to solve in this model is the nature and origin of the boiler. How was the boiler created? What is the nature of the boiler? How can we approach these questions? In spite of the difficulty to explain the genesis of the boiler, we can find a boiler at the center of a barred spiral galaxy as a luminous body. So the rotation boiler hypothesis is not a scientific fiction but real, and could be examined by astronomic observations and scientific experiments. Also, we will reveal how to create the boiler and the mechanism of its rotation later.

Here I proposed three possible hypotheses to explain the force of a spiral galaxy. Those are a universal gravity model, a tornado model, and a rotating boiler model. Each model proposes unique nature of the galactic movements. Furthermore, each model may suggest unique aspects of distribution of elements in a galaxy, unique nature of optics in a galaxy, unique mechanism for the genesis of the galaxy and its stars. Which hypothesis is right? Do all the three models have some problems? Is it possible to make the forth model, which could overcome the problems in the three hypotheses? From the next chapter, let us examine each hypothesis by the obtained scientific data available for everybody.

Chapter Four
Elements of the Galaxy

We are touched by splendor and loneliness of galaxies that bloom in a vast space of the cosmos. Billions of stars twinkle in a galaxy just like a firework at a night festival. Galaxies also appears lonely islands in the space ocean. The cosmic space around a galaxy is almost vacuum filled with quite low density of hydrogen, that is, the monad as an atomic mass unit. Let us discuss in detail on the genesis and distribution of elements in a galaxy, which exist lonely in a vast space filled by extremely low-density hydrogen.

In the twentieth century, modern science succeeded to identify ninety-two elements existing naturally on the earth. In the periodic table of the elements, we found all the natural numbers from one to ninety-two as an atomic number of each element from hydrogen to uranium. And the mass of each atom is the integral multiplication of that of hydrogen atom without any exception. Moreover, modern physics succeeded in the creation of artifact elements, whose atomic numbers are more than ninety-two. What kinds of and how many elements are there naturally in the Galaxy? Every element is created by the unification of monads or hydrogen atoms as an atomic mass unit. This law of the element should govern all the cosmos without exception because the nature of the monad is universal everywhere in the cosmos. As far as elements are created from hydrogen atoms, not only all the stars in the Galaxy but also all the galaxies in the whole cosmos should share a common periodic table of the elements. Since heavy elements could be artificially created by a sharp collision of light elements in modern physics, some artificial elements might have an atomic number of more than a hundred and ten. Therefore, some galaxies might have elements with atomic numbers of a hundred and twenties or thirties. However, modern nuclear physics reveals that all the stable elements have atomic numbers less than ninety-three. Any element with atomic number of more than ninety-two is unstable, and consequently performs spontaneous nuclear fission to form elements with low atomic numbers. Taken all together, elements in the Galaxy should have exactly the same nature as those on the earth under the law of the element.

The synthetic ratio of elements, which are created for the first time by the collision, depends on the total energy of the collision of hydrogen atoms in space. Larger energy could create heavier elements. If the total energy for creation of elements is small, the ratio of light elements may increase. Some elements naturally start to undergo fission with radioactivity just after the creation of the elements from hydrogen atoms. And the ratio of species of elements will become similar to that on the earth after billions of years. How are elements created from hydrogen

atoms?   How do hydrogen atoms collide after acceleration?
Are hydrogen atoms accelerated by the universal gravity on
themselves?   Are they accelerated with some electric force
in space?   Or do they blow and squeeze themselves on the
wind in space?   Elementary particle physics is important
to understand the genesis of the Galaxy by addressing
these questions, which are beyond the scope of this tiny
book.   Readers are strongly recommended to refer to the
technical books on particle and nuclear physics if necessary
for further study.   Here in this book, we should focus on
the genesis of the Galaxy from the standing point on the
astronomy and cosmology.

The mass distribution of the Galaxy is one of the
unsolved problems in the modern science.   Some scientists
declare that almost all the mass of the Galaxy come from
unknown substance, that is, dark matter.   Dark matter
does not belong to elements in the periodic table, and its
nature is totally unknown except for its mass.   Under this
situation, it is nonsense to discuss on the mass distribution
of the Galaxy at all.   Moreover, it may fall into scientific
fiction for us to discuss the distribution of elements in the
Galaxy.   Here, we can just conclude that the Galaxy
contains around a hundred kinds of elements, which are
created by collision and integration of accelerated hydrogen
atoms in apace, in order to form stars and planets such as
the sun and the earth.

Chapter Five
Optics of the Galaxy

In order to inspect the three hypotheses on motion of the Galaxy, we have another strategy using the optical nature of the Galaxy. To tell the truth, human beings do not yet succeed in reaching a planet outside of the solar system by a spacecraft. So, we have little technique to obtain information on the outside of the solar system other than the astronomic observation using a high-resolution telescope. It is also possible to receive an electric wave using a radio telescope. At present, experiments of the Galaxy are mainly performed by the optical observation using various types of high-resolution telescopes. Consequently, the optical data is the most reliable results to be considered for the inspection of the three hypotheses on motion of the Galaxy. In this chapter, let us examine the three hypotheses one by one using the optical data obtained in the Galaxy.

At first, let us look up at the Milky Way at the sky in the night. A beautiful Milky Way passes through the sphere in the heaven with a path of the stars. In a wide region of the path, stars appear to form a three-layer structure, in which two bright layers sandwich an internal dark layer. The wide region of the Milky Way corresponds to the central luminous body, which may be referred to the yolk of a sunny side up egg compared to the Galaxy. In the galactic sunny side up, the sun locates at the middle of the egg white that surrounds the yolk. Thus, we are just watching the yolk from the middle of the egg white when we watch the wide region of stars from the earth. And the thin path of stars is comparable to the egg white of the sunny side up.

We can see light with wavelength between three hundred and eight hundred nanometers by naked eyes. Visible light is usually radiated from a black body at around six thousand centigrade degrees on its surface. A black body at less than a thousand degrees radiates infrared and far infrared rays, which we could not see by our naked eyes. And a black body at more than six thousands degrees also radiates visible lights together with ultraviolet and X-rays with very short wavelength, which we could not see by our naked eyes, either. The ultraviolet and X-rays do not reach at the surface of the earth because those lights are absorbed by the atmosphere on the earth. As a consequence, we can see the distribution of subjects at more than a thousand centigrade degrees in the Galaxy as the visible Milky Way in the sky.

Why does the middle layer of the wide region appear dark? Is the dark layer real empty? Or is there something dark such as clouds, which conceal the existing stars, on a line of the long axis of the central elliptic body? In other words, the question is whether invisible clouds exist or not at the dark layer of the central luminous body of the Galaxy. Since materials at around a thousand degrees radiate infrared rays, astronomers can easily answer the above questions using a telescope equipped with an infrared detection device. And the data clearly demonstrated that the dark layer emits infrared light so brightly. In other words, the center of the Galaxy is not an empty space but filled with invisible gaseous clouds at less than a thousand degrees. The invisible gaseous materials should mask the light radiating from shining stars among and behind them.

Although the luminance of the gaseous clouds is lower than that of the stars, the total mass of the gaseous clouds is enormously larger than that of the stars because gaseous clouds fill almost all the space in the dark region of the central body of the Galaxy while the twinkling stars occupy only tiny spots in a vast space. It is quite likely that the stars are born from the gaseous clouds by condensation just like rain drops in the clouds. Since the condensed liquid is cooler than the gas in the clouds, the baby liquid star should not radiate visible lights. How do the baby cool stars become luminescent after condensation? The only way for the stars to become hot should be the generation of heat by themselves. Since the gaseous clouds contain all the elements including radioactive heavy metals such as uranium, the distilled and condensed baby stars may naturally concentrate those radioactive materials at the core and accelerate nuclear fission just like a nuclear plant for electricity.   Now, the baby cool star starts to produce heat by itself from inside, and will become as hot as six thousands centigrade degrees like the sun.   Because of the huge mass of a star, the surface pressure of the atmosphere becomes quite high enough to keep the hot star in a solid or liquid phase as a spherical body in space.   If this speculation is right, the primary gaseous clouds of the mixed elements, which expand from center to periphery of the Galaxy in a spiral way, are condensed to create stars by radiation cooling at the peripheral region in the Galaxy. This is consistent with the rotating boiler model.

The Galaxy radiates gamma rays, X-rays, and line spectra together with the visible and infrared rays mentioned above.   Both the gamma and X-rays are radiated from the center of the Galaxy upwardly and downwardly on the vortex axis.   Since gamma rays are radiated from radioactive materials, it is quite likely that some nuclear reactions of radioactive materials occur at the vortex center of the Galaxy.   Since X-rays are often emitted from an electric discharge tube, it is possible that the Galaxy performs a large-scale discharge at the center of its vortex.   The electric discharge could be generated by beta decay of radioactive materials.   Taken all together, it is quite likely that the Galaxy performs conversion of elements at the center of the vortex.   It could be a nuclear fusion of hydrogen atoms to create heavy elements, and it could be a nuclear fission of radioactive heavy elements, and could be both.

The motion of materials around the center of the Galaxy is quite complicated in a chaotic situation, and some air currents blow into the center of the vortex, while others blow up and down from the center along the vortex axis, and the others spout out radially from the center of the Galaxy within a plane of the vortex. All the air currents may be mixed up around the center. Let us consider the precise motion around the center in the next chapter.

Chapter Six
Genesis of the Galaxy

Ninety percent of atoms in the solar system are hydrogen, and ten percent of atoms in the solar system are helium. All the other elements correspond to less than one percent of total atoms in the solar system. If we convert the number of atoms into mass, hydrogen is seventy percent, helium is thirty percent, and all the other elements are less than two percent of the total mass of the solar system. The composition of elements of the solar system is considered to be rich in heavy elements when compared to that of the outer space, suggesting that almost all the atoms in space are hydrogen. How is a galaxy born in such a vast space filled with quite low density of hydrogen?

At first, let us discuss the possibility that a part of space, which is filled with hydrogen gas, start to accumulate into a center of mass by the universal gravity after the shake of balance of power among hydrogen atoms. Hydrogen gas exists as a hydrogen molecule consisting of two hydrogen atoms. The hydrogen gas behaves under the law of the ideal gas. As a result, the gas does not accumulate into a point by the universal gravity. This is also true for all the gas consisting of any elements in the periodic table. Is it possible for liquid or solid materials, which are drifting separately in a space to concentrate, into one body by the universal gravity of themselves? Yes, it is. Especially, heavy materials such as meteorites and asteroids have a large power to attract other materials around it by the universal gravity, although an attractive force between two bodies causes an elliptic motion rather than collision of the two. However, the friction between the bodies and surrounding media reduces the angular momentum of the bodies until the two bodies collide. By the universal gravity with friction against its medium gas, the cosmic dusts grow gradually to become meteorites, planets, and stars, which might start to twinkle with converting the original potential energy to the heat. The twinkling stars also accumulate to become a huge aggregate, which increases its density more and more by enormous pressure of its universal gravity. Finally, the huge aggregate may induce internal collapse and become a black hole, which absorbs everything around in the galaxy. Some astronomers insist that the absorption of everything by a black hole results in vortex motion of stars, which we see as a spiral galaxy in the cosmos.

However, this model contains two serious problems. One problem is the friction of the medium gas. If there is no friction in a perfect vacuum, almost all the independent bodies moves on an elliptic orbital under the control of the universal gravity. In this case, we take all the quadratic curves into the elliptic orbital. In the elliptic orbitals, the two bodies do not collapse to each other with only one exception, which is a linear movement with zero angular momentum around the center of mass of the two bodies. Since we could not detect any linear or elliptic motion of stars in a galaxy, the universal gravity model requires a friction of the medium gas to produce a spiral shape of the galaxy. In a medium gas with frictional resistance, the cosmic dusts and solid materials move slower and slower gradually as time goes by, and finally everything falls down onto the center of the spiral by the universal gravity. On the other hand, the frictional effect of the medium gas causes the hydrodynamic flow of the entire physical system of stars. In other words, the spiral motion of solid bodies should be spouted out asymmetrically in one direction by a space wind. Therefore, cosmic dusts do not form a symmetric spiral by the universal gravity in a frictional medium that blows in a space. Theoretically in the universal gravity model, a symmetric spiral galaxy should be formed only in a calm space without wind. If cosmic winds flash and disturb the galaxies, they will form unique shapes with elongation to one direction just like a marbling print. This speculation is obviously inconsistent with astronomic data of galaxies obtained by a modern high-resolution telescope. Therefore, the universal gravity model clearly indicates that cosmic dusts perform vortex motion in a spiral of a galaxy only in a calm space without any wind.

The other problem of the universal gravity model is the

difficultness to explain the mechanism how the spiral is restricted in a plane. As we discussed above, materials fall down in a spiral manner through a frictional atmosphere by the universal gravity, and everything is finally absorbed into the center of mass or a black hole. A spiral belongs to a quadratic curve, which exists with in a single plane containing the center of mass and the velocity vector of the body. Since the direction of velocity vectors are expected random in a calm space, a lot of bodies in a galaxy should form a three-dimensional spiral. However, this is not the case in the real galaxy. Almost all spiral galaxies in the cosmos belong to the two-dimensional spiral. If the universal gravity model is right, and if the observed two-dimensional spiral galaxy is true, then almost all the materials in the Galaxy originally have velocity vectors within a single plane. This does not happen in a calm atmosphere at all. So, we conclude that the universal gravity model does not work as model for the motion of the Galaxy.

Then, how was the Galaxy created in space? It is possible that a tornado, which is caused by space wind of cosmic atmosphere, created the Galaxy. When a gust of wind pass in a vast space, there must be many whirlwinds or tornados here and there along the path of the wind. The space tornado of hydrogen gas is the first cry of a new baby of the Galaxy. It is quite natural for the whirlwind of a tornado to cause a spiral of the Galaxy, in which

28

everything distributes in a spiral shape and swirls around the center of the spiral. Materials are caught up in the maelstrom and some of them may escape from the cyclone storm upward or downward along the axis of the spiral. Since a whirlwind of the space tornado rules both the distribution and motion of all the materials, the Galaxy spirals naturally.

In a vast space, there are some lonely galaxies, which might be created by a small whirlwind. However, several galaxies gather at some region in a space frequently. In such a region, it might be happen that a gust of storm blew in a cosmic space creating a series of tornados along the path of wind at the same time. The group of galaxies is called a cluster. This could be compared to maelstroms in the Lofoten islands off the Norwegian coast. The hypothesis that a space tornado creates a galaxy explains not only the genesis of a single galaxy but also the genesis of clusters of galaxies in the whole cosmos. The Galaxy or our Milky Way forms a cluster with neighbor galaxies, Large and Small Magellan Clouds. Therefore, We can speculate an atmospheric current of space hydrogen gas from the distribution of galaxies in the cosmos using the space tornado model. Once we identify the space hydrogen gas currents, it is possible to expect the place and timing of the birth of galaxies in the future.

After the first cry of a baby galaxy created by a space tornado, the hydrogen gas is squeezed into a center of the maelstrom by the pressure of the whirlwind. The hydrogen gas pours into the center of the spiral from all directions, and becomes high-pressure, high-temperature, and high-density by the adiabatic compression. A total energy of a tornado could be huge enough for creation of the Galaxy from the hydrogen gas. What's happen in the agitated hydrogen gas under high-pressure at high-temperature, with high-density? Now, we should remember the optic nature of the Galaxy in chapter five. The center of the spiral radiates beta and gamma rays, suggesting that nuclear reaction occurs there. However, the only materials pored into the center is hydrogen gas, which could not perform nuclear fission. Although we could not rule out the possibility that the hydrogen gas contains small amounts of radioactive cosmic dusts, the amount of radioactive dusts is quite small in space and is not enough to create the entire Galaxy. Then, what's really happen in the center of the spiral? It is quite likely that nuclear fusion of hydrogen atoms creates elements such as helium there. Only helium? The created helium gas itself also agitated with hydrogen at high-temperature under high-pressure, with high density in the space tornado. It will create lithium, beryllium, boron, carbon, nitrogen, fluorine, and neon. Moreover, it must create uranium. Furthermore, it will create super heavy elements with atomic number of more than hundred, which decay by the nuclear fission immediately after their creation.   This must be the source of the beta and gamma rays radiated from the center of the spiral Galaxy.   In short, the center of cosmic tornado is the synthesizer of elements for creation of the Galaxy from the space hydrogen gas.   In this way, the space tornado delivers a baby galaxy crying for the first time in space by synthesis of all elements required for the galaxy from hydrogen gas. This is the moment of the birth of a galaxy.   And this is also the moment of the birth of elements.   Now, there is a primary chaos of all the elements mixed in the vortex manner under high pressure at high temperature with high density in a gaseous phase.   This is a wonderful event in the cosmos, and in Japan we clap our hands in prayer for celebration of the birth as a traditional custom.

Chapter Seven
Genesis of Stars

The primary chaos of the galaxy, which is created by a space tornado, grows gradually and becomes ready to produce stars. A story of the birth of the sun in the solar system has been written in my book titled "The Sun" in detail. So, here we focus on the story from the primary stage of the vortex vapor gas to the month in which star birth is expected in the spiral galaxy.

The primary vapor gas was created from hydrogen gas by nuclear fusion at the center of a space tornado. The law of ideal gas rules both the hydrogen gas and the primary mixed gas. The molecular mass of hydrogen gas is two, while those of helium, argon, and xenon are four, forty, and one hundred thirty one, respectively. Consequently, the volume of the hydrogen gas reduces after a nuclear fusion reaction. In other words, the hydrogen gas increases its density at a moment of the transmutation of elements at the center of the vortex in the Galaxy. Thereafter, the nuclear fusion gas spouts out at a high density from the center of the Galaxy.

How does the high-density gas spout out from the center of the Galaxy? Previously, we already discussed the wind of hydrogen gas spouting out from a space tornado. The hydrogen gas spouts out upwardly or downwardly along the rotation axis of the spiral. How about the primary mixed gas in high-density?   Does the gas spout from the center of the vortex upwardly or downwardly along the rotation axis like the hydrogen gas?   Since the hydrogen gas pours into the center of the whirlwind in a plane of the spiral, the synthesized mixed gas should spout out on an air current from the high pressure center to the low pressure upward or downward along the rotation axis.   Because of the high speed of rotation, larger centrifugal force acts on the high-density gasses of heavier elements in the mixed gas at the center of the tornado.   Consequently, the high-density heavy-element gas should start to expand radially from the spiral axis just after the gas spouts out upward or downward from the whirlwind.   You may imagine the rain spray flying from the rotating umbrella through the air. You may also imagine the real whirlwind or tornado on the earth that scatters everything away all over the ground. It might be possible that an elliptic galaxy is transformed to spiral galaxy when the high-density gas starts to expand radially from the spiral axis of the tornado.   Or it is also possible that an elliptic galaxy is created by a tiny whirlwind, which is enough to produce radioactive elements for a central luminous body but not enough to scatter the high-density gas away from the ellipse.   In the spiral galaxy, heavy element gasses such as uranium gas will expand radially and may be concentrated close to the spiral plane by the centrifugal power.   On the contrary, light element gasses such as hydrogen and helium gasses will spout out upwardly and downwardly along the spiral axis for the great distance. In short, a space tornado swirls and accumulates a hydrogen gas at the center of the spiral, creates all the elements required for the galaxy by nuclear fusion, and spouts out the light and heavy element gasses vertically and horizontally to the spiral plane, respectively.

It is not so difficult to imagine the motion of the hydrogen and light element gasses spout out from the center of the whirlwind vertically above and below the spiral plane. The gasses will expand just like smoke emitting from a volcano. How do the relatively heavy gasses behave after spouting out radially from the center of the spiral? The gasses expand by two different forces, that is, the high pressure of the gasses compressed at the center of vortex and the centrifugal force of the rotating heavy element gasses at high density. If a tornado has already blown off, the heavy element gasses will expand radially or on a straight way from the center. Since the central body still keep rotating around itself by the law of inertia, the shape of the spouting gas becomes spiral. This is nothing other than the rotation boiler model itself. If a tornado still swirls the hydrogen gas, spiral arms of the heavy element gas in the galaxy drifts on the current of the hydrogen gas in the whirlwind. Anyway, the high-temperature, high-pressure, and high-density heavy element gasses condense to become liquid droplets of heavy elements by adiabatic expansion and radiation cooling in the spiral arms.   Then, the liquid droplet congeals to form a solid spherical body with radioactive elements inside, which heat up the body to more than thousands centigrade degrees for twinkling by the nuclear fission.   This is the secret story on the birth of stars twinkling in a mother galaxy.

We admire the beauty of stars twinkling in the sky at night.   A spiral galaxy is really a cosmic art with a lot of twinkling stars, and it is a gorgeous firework flowering at the night sky.   It should also be noted that the fireworks are fruits of invisible space tornados produced by a thousand winds that blow in the cosmos.

Chapter Eight
Life of Stars

The heavy element gasses, which have been synthesized by the nuclear fission at the center of the spiral galaxy, create stars by condensation during the expansion in the spiral disk by the centrifugal force.   Here, let us consider the whole life of stars twinkling in a spiral galaxy.   Since the heavy element gasses expand radially in a spiral plane by the centrifugal force, it is no doubt that the gasses moves from inside to outside.   As traveling in the galactic disk, the dense fog of heavy element gasses in chaos condenses to become stars by radiation cooling and consequently the atmosphere becomes clear up.   In other words, the fog of mixed elements in chaos creates stars and planets together with clear sky in heaven.   Then, a star starts twinkling as a luminous body at high temperature heated by nuclear fission energy of radioactive materials, which was concentrated at its core by a distillation process. Now, stars twinkle brilliantly in the sky.   Then, stars keep on travelling toward outside of the galactic disc and gradually decay and finally reduce back to the hydrogen gas outside the galaxy.

Now, we found that almost all the elements on the earth are stable isotopes, whose life span is more than tens of billions years, since the earth was born five billion years ago.   All the elements, however, should decay and collapse someday in the future. The sun is not the exception, and the solar system will return back to the hydrogen gas.

A twinkling star is a stable power plant of nuclear fission reaction using radioactive materials such as uranium for energy source. Once the nuclear fission reaction runs away in a reckless manner, the star will explode suddenly to become a supernova. Stably twinkling stars will spent its nuclear energy during their life and gradually lose the brightness. Finally, stars finish its life and go back to a thousand winds of hydrogen that blow in space forever.

Chapter Nine
Life of the Galaxy

According to the space tornado model of the Galaxy, we have discussed a life of a galaxy from its birth to the whole life of stars based on the scientific data obtained by modern astronomers. Here in this chapter, let us look through the whole life of a galaxy.

In the beginning, there is nothing but an infinite space filled with a hydrogen gas. In a vast space filled with an infinite hydrogen gas, a thousand winds blow occasionally here and there with a great number of whirlwinds around the stream. In the whirlwind, hydrogen gas pores into a center of the vortex to form a space tornado spiraling vigorously at high temperature under high pressure in a high density. At the center of the tornado, nuclear fusion of hydrogen happens and elements are created as a mixed vortex gasses in chaos. Among the elements, relatively light elements such as hydrogen and helium gasses are spouted out upwardly and downwardly along the spiral axis of the tornado by the high pressure of gasses, while relatively heavy elements such as uranium gasses are spouted out radially in the spiral plane of the tornado by the centrifugal force. The primary chaotic gas of heavy elements travels toward the outside of the galaxy to form a disc, in which the gas condensed to produce stars by adiabatic expansion and radiation cooling. Accordingly the chaotic fog of the gas clears up in the sky. Since some radioactive heavy elements are separated and condensed at a core of the spherical body of a star by distillation, the star naturally starts nuclear fission to heat up its surface for emission of light. The twinkling stars spread and rotate riding on the whirlwind to create a large-blossomed firework with a rotating spiral in the cosmos. Later on, the galaxy uses up the energy of the radioactive elements and the twinkle of the stars slacken off. Finally, even stable elements collapse and reduce back to a hydrogen gas, which fills calmly in space after the storm is gone. Now, there is another gust of wind blowing somewhere in the cosmos to produce tornados. In this way, we have a thousand winds that blow forever here and there in space, where any number of galaxies are born, grown, and gone at any number of times.

Chapter Ten
Life in the Galaxy

When we see the Milky Way at the night sky, we feel to be human beings standing in front of the cosmos. We are so tiny living things when compared to the whole cosmos. We are just one kind of mammals, which appeared only a million years ago on the surface of the earth. In our body, however, amazing and highly sophisticated macromolecules are synthesized and play a role in life activities. It is really miracle that a huge number of atoms are assembled as functional complex macromolecules such as genes, enzymes, and biological membranes to perform a beautiful and wonderful organic life activities. Who created human beings? No one could match the super talent of the creation, which is far beyond the human knowledge and power. Although our body appears so small as a tiny dust on a surface of a planet when compared to the cosmos, our macromolecule contains super sophisticated information as high as any molecule in the Galaxy. In the solar system, human beings sing the praises of wisdom and play a role as a unique living thins that have advanced intelligence. I previously wrote the history on the creation of living thins from elements in a book "Life" from a standing point of the modern molecular biology. In short, the living things are created orderly from a limited number of elements in the periodic table. The law of the life should rule all the living things in the cosmos, and consequently living things could be created anytime and anywhere in space once the condition is prepared. Therefore, it is no doubt that a lot of living things are created on planets outside of the solar system in the Galaxy. It should be happen in some star systems at the peripheral region in a disc of the Galaxy. Since stars will travel and evolve from the center to outside of the Galaxy, more intelligent living things will exist on a planet of a star system at more peripheral region of the Galaxy. Nobody knows whether they already travel much faster than light in the Galaxy or not. However, I believe that they already do.

In order to travel in the Galaxy, we definitely need to move at a speed faster than that of light. Can we travel faster than light? How can it be possible? What is light? To tell the truth, the modern scientists still argue about those fundamental questions everyday. In this book, we have discussed the Galaxy, which is huge phenomena in the cosmos, from a standing point of natural philosophy. However, we now recognize it quite important to reveal secrets of particle physics such as light, the universal gravity, electricity, magnets, temperature, radio wave, an atom, and monad or an atomic mass unit, in order to find out the way for human beings to go on. If we could join the social of intellectual living things in the Galaxy after we solve all the questions above, we need to have the responsibility, morals, personality, love, gentleness, and culture as cosmopolitans to manage the Galaxy. Since human beings already have an atomic power in the twentieth century, it is a good idea to start preparing for the morals of our behavior, which we need to master before the further advance of science and technology.

Finally, I pray for happy future of human beings in the Galaxy.

Reference Books

1. The Realm of the Nebulae, by Edwin Hubble (1936), published from Yale University Press, Newhaven
2. Life, by Hiroyuki Aizawa (2015), published from Aizawa Science Museum Press, Kasukabe
3. The Sun, by Hiroyuki Aizawa (2015), published from Aizawa Science Museum Press, Kasukabe

An Epilogue

A gust of wind gives rise to a storm of whirlwind here and there in space, and clusters of galaxies are born, grown, bloom as a beautiful firework, and gone heroically. Fallen petals return back to the hydrogen gas in heaven, and there are just blowing winds and nothing else. A thousand winds of hydrogen blow in a vast space of the cosmos forever. Also, living things are created during a history of a galaxy, and develop the intelligent ability to perform highly sophisticated social culture by the evolution. That is the world we live at present.

Once we aware this truth of the world, we suddenly understand that we are created and saved by mighty power, that we need to keep and develop our social culture and life, that we are a member of living things in the Galaxy, and that we are expected to spend our life just as human beings. That is the most important, and there is nothing else other than that, and that is enough. The modern cultural society of human beings is one of great fruits of the long history of development of the Galaxy. Nobody knows how great it could bloom further from now on. It is destiny and pleasure for the Galaxy to bloom as great as possible in the cosmos.

It is a great luxury to spend our life in communion with nature in the Galaxy. It is a great pleasure to reveal that to love nature is exactly to love ourselves because the whirlwind in the cosmos is our ultimate birthplace and nature is mother of us. It is quite lucky to have a birth in such a good age on the earth. It is important to keep on living everyday, to live to cherish, to cultivate the sense of molarity, and to take care of life with love. It is my great pleasure to share this truth and hope with as many readers as possible.

Finally, I would like to express my special thanks to my family, Yoko, Kodai, Arisa, and Hiroto for supporting my life and writing this book.

At home in Kasukabe
July 27th, 2015
Hiroyuki Aizawa

www.ingramcontent.com/pod-product-compliance
Lightning Source LLC
Chambersburg PA
CBHW071013180526
45168CB00003B/1409